U.S. ARMED FORCES

U.S. NATIONAL GUARD

by Allan Morey

po**go**

Ideas for Parents and Teachers

Pogo Books let children practice reading informational text while introducing them to nonfiction features such as headings, labels, sidebars, maps, and diagrams, as well as a table of contents, glossary, and index.

Carefully leveled text with a strong photo match offers early fluent readers the support they need to succeed.

Before Reading

- "Walk" through the book and point out the various nonfiction features. Ask the student what purpose each feature serves.
- Look at the glossary together. Read and discuss the words.

Read the Book

- Have the child read the book independently.
- Invite him or her to list questions that arise from reading.

After Reading

- Discuss the child's questions. Talk about how he or she might find answers to those questions.
- Prompt the child to think more. Ask: Before reading this book, did you know what types of jobs U.S. National Guard members perform? What more would you like to learn about the U.S. National Guard?

Pogo Books are published by Jump!
5357 Penn Avenue South
Minneapolis, MN 55419
www.jumplibrary.com

Library of Congress Cataloging-in-Publication Data

Names: Morey, Allan, author.
Title: U.S. National Guard / by Allan Morey.
Description: Minneapolis, MN: Jump!, 2021.
Series: U.S. Armed Forces
Includes index. | Audience: Ages 7-10
Identifiers: LCCN 2019050583 (print)
LCCN 2019050584 (ebook)
ISBN 9781645274278 (hardcover)
ISBN 9781645274285 (ebook)
Subjects: LCSH: United States—National Guard
Juvenile literature.
Classification: LCC UA42 .M67 2021 (print)
LCC UA42 (ebook) | DDC 355.3/70973—dc23
LC record available at https://lccn.loc.gov/2019050583
LC ebook record available at https://lccn.loc.gov/2019050584

Editor: Susanne Bushman
Designer: Molly Ballanger

Content Consultant: Jennifer Rechtfertig, Air National Guard

Photo Credits: Scott Olson/Getty, cover; Airman 1st Class Kevin Donaldson, 1 (foreground); turtix/Shutterstock, 1 (background); Master Sgt. Mark C. Olsen/U.S. Air Force, 3; SSG Robert Adams, 4; Staff Sgt. Daniel J. Martinez/Air National Guard, 5; Staff Sgt. Michael Giles/U.S. Army, 6-7t; Randy Burlingame/U.S. Air National Guard, 6-7b; Sgt. Brian Calhoun, 8-9; Staff Sgt. Michael Broughey, 10; Edwin L. Wriston/U.S. Army National Guard, 11, 20-21br; Staff Sgt. Osvaldo Equite/U.S. Army, 12-13; SDI Productions/iStock, 14-15; Senior Airman Cody Martin/U.S. Air National Guard, 16-17; 1st Lt. Leland White/U.S. Army National Guard, 18; Master Sgt. David Loeffler, 19; Staff Sgt. Balinda O'Neal Dresel/U.S. Army National Guard, 20-21tl; Staff Sgt. Alex Baum/Wisconsin National Guard, 20-21tr; Spc. Joseph K. VonNida/U.S. Army National Guard, 20-21bl; Senior Julia Santiago/U.S. Air National Guard, 23.

Printed in the United States of America at Corporate Graphics in North Mankato, Minnesota.

TABLE OF CONTENTS

CHAPTER 1

TWO GROUPS OF GUARDS

People in uniforms fill sandbags. Why? They are building a wall. It will hold back floodwaters. Who are these people? They are members of the U.S. National Guard.

sandbags

Guard members help after **natural disasters**. How? They clean up after tornadoes. They rescue people after floods, earthquakes, and hurricanes.

members of the
U.S. Army National Guard

members of the
U.S. Air National Guard

The U.S. National Guard is part of the U.S. military. It has two parts. One is the U.S. Army National Guard. The other is the U.S. Air National Guard. Each of the 50 states has both. During times of peace, each member reports to his or her state's **governor**.

DID YOU KNOW?

Who else commands the Guard? The U.S. president. He or she decides what the U.S. military will do during war and national emergencies.

Guard members help **branches** of the U.S. military. During **conflicts**, the Air National Guard supports the U.S. Air Force. The Army National Guard helps the U.S. Army. The Air Force might help the Army National Guard, too. How? They **transport** tanks.

tank

TAKE A LOOK!

Compare the number of members in each branch.
Which has the most? Which has the least?

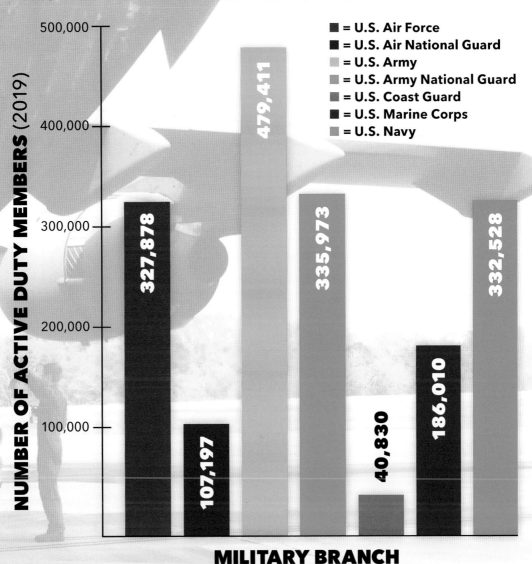

NUMBER OF ACTIVE DUTY MEMBERS (2019)

500,000

400,000

300,000

200,000

100,000

- ■ = U.S. Air Force
- ■ = U.S. Air National Guard
- ■ = U.S. Army
- ■ = U.S. Army National Guard
- ■ = U.S. Coast Guard
- ■ = U.S. Marine Corps
- ■ = U.S. Navy

327,878

107,197

479,411

335,973

40,830

186,010

332,528

MILITARY BRANCH

CHAPTER 2

TRAINING AND JOBS

Recruits go to Basic Training. They must pass physical tests.

They go through skills training, too. Like what? They learn to fire weapons.

Some Army National Guard members train to be Green Berets. This is a **special force**. They get **deployed** once every two to three years. They might train other countries' armies. Or they look for **terrorists**.

WHAT DO YOU THINK?

National Guard Green Berets are deployed for up to 15 months. They are away from friends and family. How do you think this would feel?

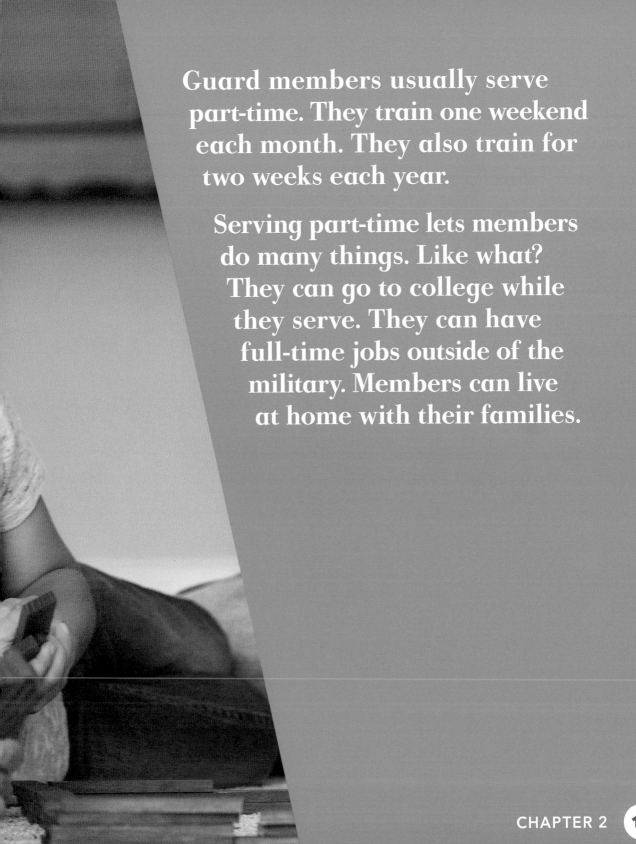

Guard members usually serve part-time. They train one weekend each month. They also train for two weeks each year.

Serving part-time lets members do many things. Like what? They can go to college while they serve. They can have full-time jobs outside of the military. Members can live at home with their families.

What military jobs can Guard members do? Some train to fix vehicles. Some work on computers. Others learn to **forecast** weather. This information can help during **missions**.

WHAT DO YOU THINK?

What if you joined the military? What job would you like to train to do? Why?

CHAPTER 3

NATIONAL GUARD MISSIONS

The Guard performs many different missions. In 2003, the United States entered the Iraq War (2003–2011). Army National Guard members served in **combat**.

Guard members act fast in emergencies. How? During forest fires, they fly helicopters. They drop water onto the fires.

The Guard serves an important role in the U.S. military. Members protect their state and their country. They use many different vehicles to rescue people around the world. Would you like to join the National Guard?

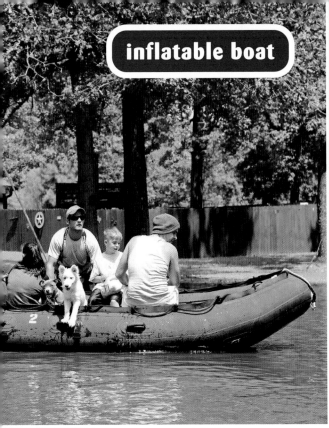

inflatable boat

high-water vehicle

small unit support vehicle

UH-60M Blackhawk helicopter

QUICK FACTS & TOOLS

TIMELINE

1947
The Air Force separates from the Army. The Air National Guard is formed.

1636
The colony of Massachusetts organizes a militia. These men are the first members of what will become the Army National Guard.

1921
The U.S. Army establishes what is now known as the 109th Airlift Squadron as the National Guard's first flying unit.

1775
The Continental Army is formed during the Revolutionary War (1775–1783).

1903
The Militia Act of 1903 creates the rules of the current Army National Guard.

U.S. NATIONAL GUARD MISSION

The federal mission of the U.S. National Guard is to maintain well-trained, well-equipped units available for prompt mobilization for war, national emergency, or as otherwise needed.

The state mission of the Air National Guard is to provide protection of life and property and to preserve peace, order, and public safety.

The state mission of the Army National Guard is to respond to battle fires, help communities deal with floods, tornadoes, hurricanes, snowstorms, or other emergency situations, and be ready to respond in times of civil unrest.

U.S. AIR NATIONAL GUARD MEMBERS:
around 107,000 (as of 2019)

U.S. ARMY NATIONAL GUARD MEMBERS:
around 335,000 (as of 2019)

branches: The groups of the U.S. military, including the U.S. Air Force, U.S. Army, U.S. Coast Guard, U.S. Marine Corps, and U.S. Navy.

combat: Fighting.

conflicts: Wars or other fights.

deployed: Sent into action.

forecast: To predict.

governor: The highest elected official of a U.S. state.

missions: Tasks or jobs.

natural disasters: Events in nature, such as hurricanes, earthquakes, and floods, that cause a lot of damage.

recruits: New members of a military force.

special force: A specialized military group that is trained to do particular tasks.

terrorists: People who use violence and threats in order to frighten people, obtain power, or force governments to do things.

transport: To carry from place to place.

INDEX

TO LEARN MORE

Finding more information is as easy as 1, 2, 3.

❶ Go to www.factsurfer.com

❷ Enter "U.S.NationalGuard" into the search box.

❸ Click the "Surf" button to see a list of websites.

FACT SURFER